KS2
Success

Age 7-11

Maths

Tests

Trevor
Dixon

Contents

(pull-out section at the back of the book)

Arithmetic

- You have 20 minutes to complete this test.
- Calculator <u>not</u> allowed.
- Use the spaces provided for your workings. Where two marks are available, you may be awarded a mark for your workings.

1 $479 + 100 =$

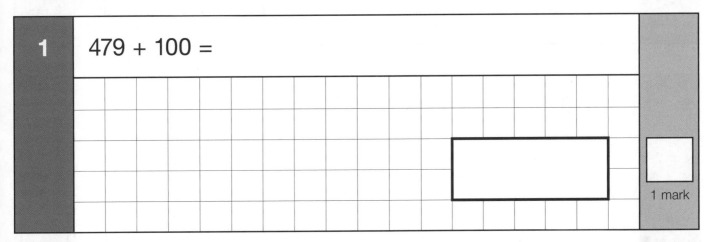

1 mark

2 $7256 + 847 =$

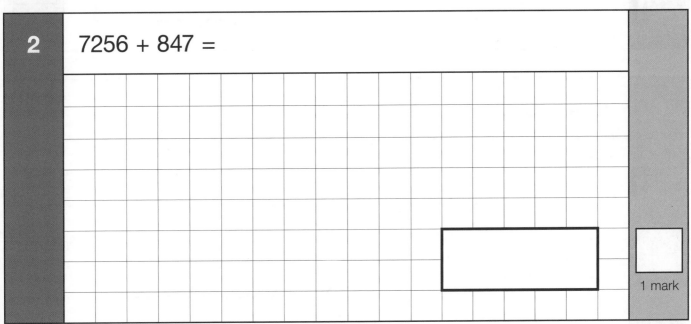

1 mark

3 $\dfrac{9}{10} + \dfrac{7}{10} =$

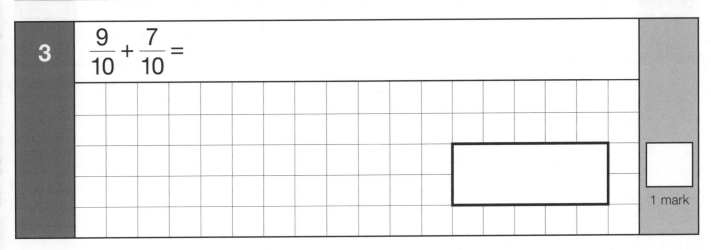

1 mark

4 594 + 10 =

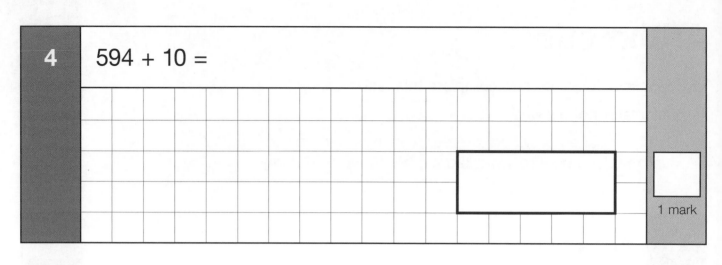

1 mark

5 14 × 6 =

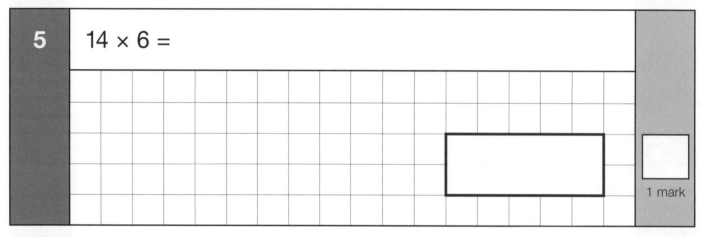

1 mark

6 3074 × 6 =

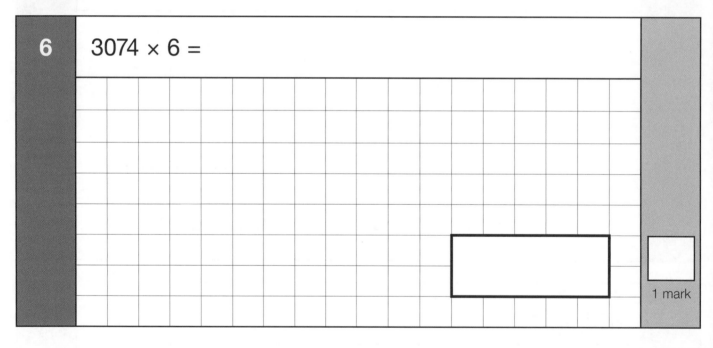

1 mark

7

$2\frac{3}{5} = \frac{\square}{5}$

1 mark

8

$\frac{7}{12} - \frac{1}{6} =$

1 mark

9

$5^2 + 5^2 =$

1 mark

10

```
    3 2 8
  ×   2 6
  ─────────
```

Show your working

2 marks

11

```
2 5 │ 1 7 7 5
```

Show your working

2 marks

6

12   6 × (46 − 29) =

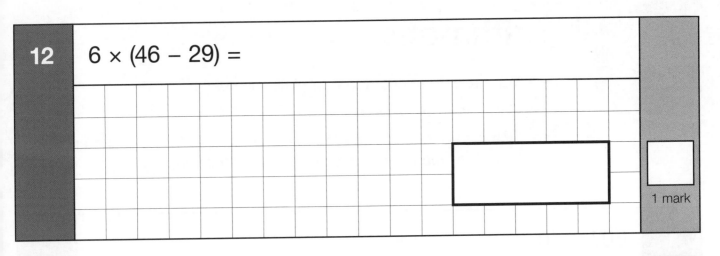

1 mark

13   12 − 20 =

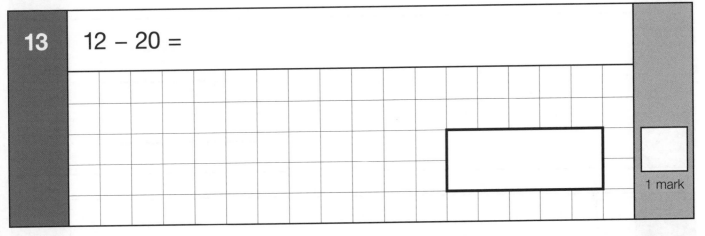

1 mark

14 $\dfrac{1}{3} \div 5 =$

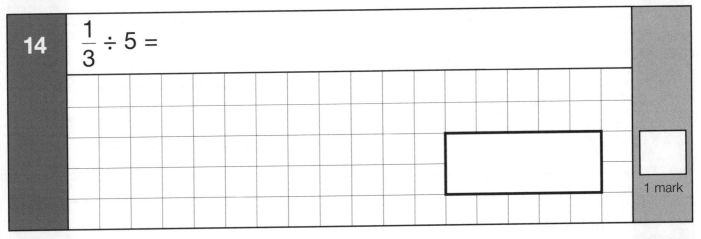

1 mark

15   0.07 × 7 =

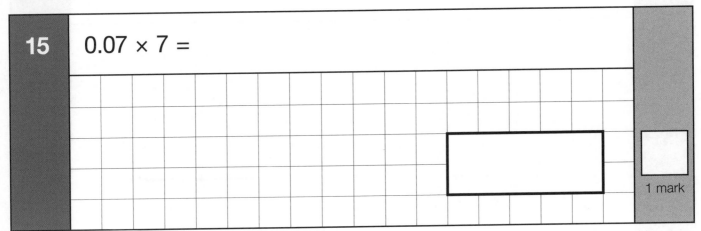

1 mark

Total _____ / 17 marks

Arithmetic

- You have 20 minutes to complete this test.
- Calculator <u>not</u> allowed.
- Use the spaces provided for your workings. Where two marks are available, you may be awarded a mark for your workings.

1 607 − 10 =

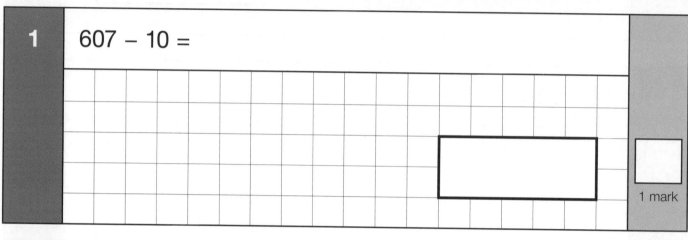

1 mark

2 $\dfrac{\square}{4} = \dfrac{1}{2}$

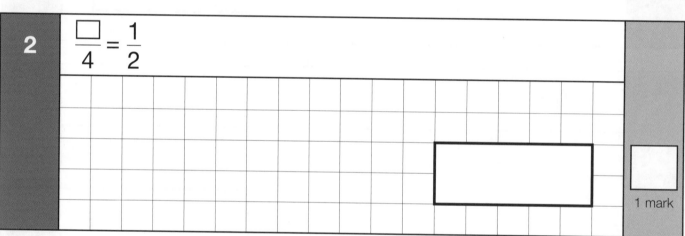

1 mark

3 648 × 6 =

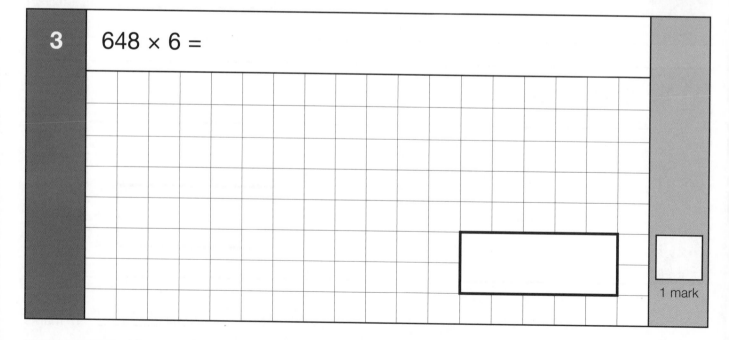

1 mark

4 $5476 \div 4 =$

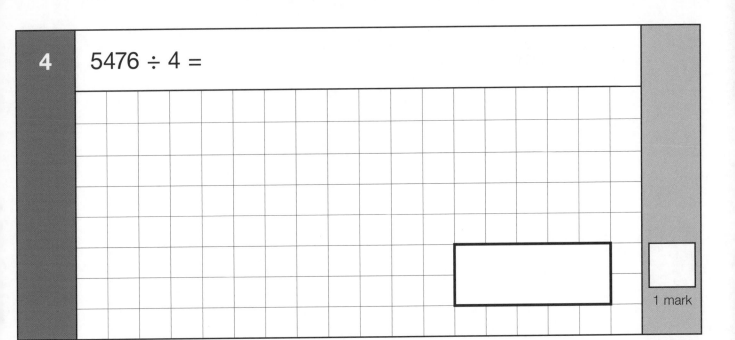

1 mark

5 $0.703 = \dfrac{7}{10} + \dfrac{3}{\square}$

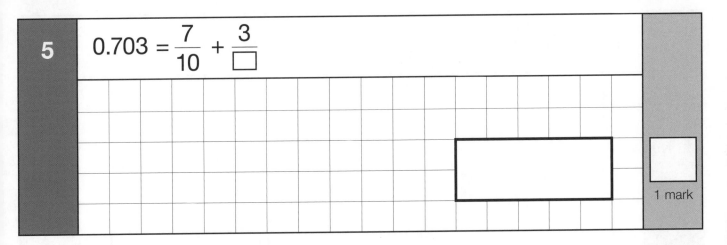

1 mark

6 $60 + 130 + 2300 =$

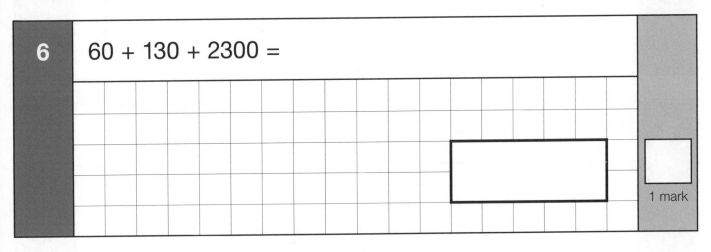

1 mark

7 7363 − 3661 =

1 mark

8 6.7 × 100 =

1 mark

9 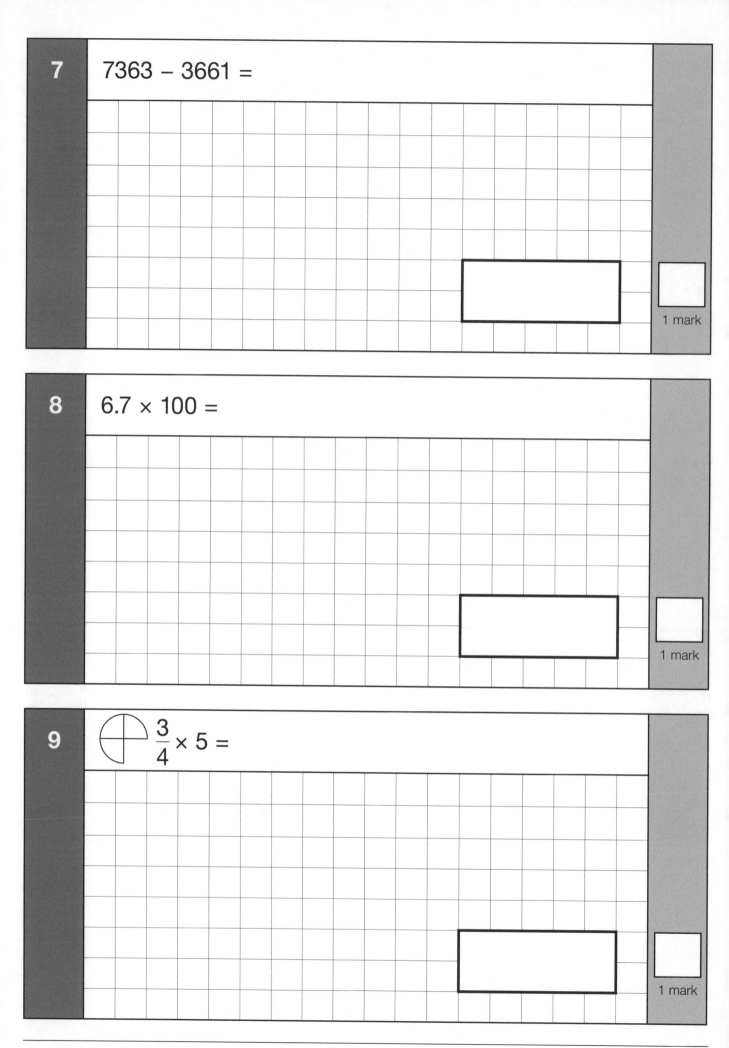 $\frac{3}{4} \times 5 =$

1 mark

10   8 − 25 =

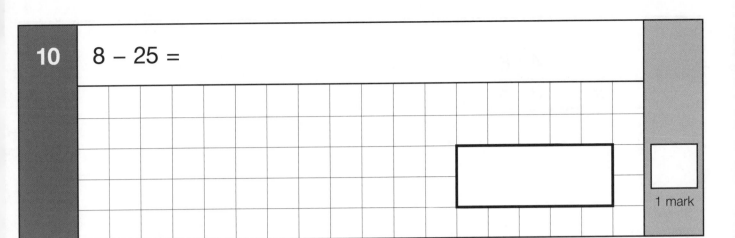

1 mark

11   560 × (305 − 297) =

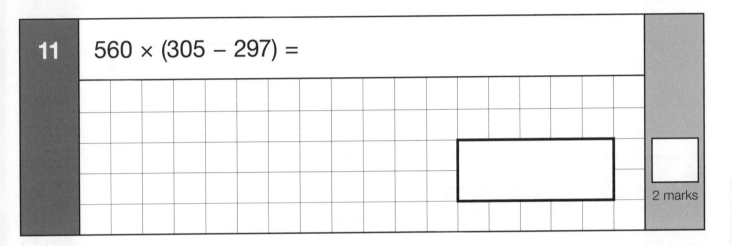

2 marks

12 $\dfrac{16}{24} = \dfrac{2}{\square}$

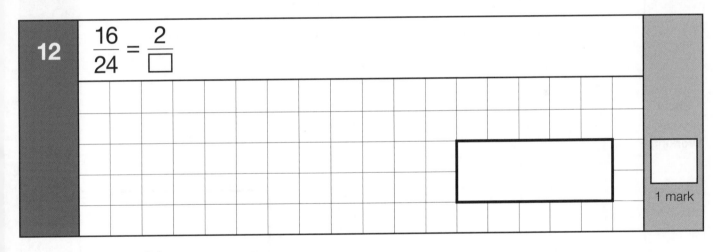

1 mark

13

Show your working

$$\begin{array}{r} 6\ 4\ 8 \\ \times\quad 3\ 5 \\ \hline \end{array}$$

2 marks

14

Show your working

$$3\ 2\ |\ 5\ 7\ 9\ 2$$

2 marks

Arithmetic

- You have 20 minutes to complete this test.
- Calculator not allowed.
- Use the spaces provided for your workings. Where two marks are available, you may be awarded a mark for your workings.

1   35.7 + 45.8 =

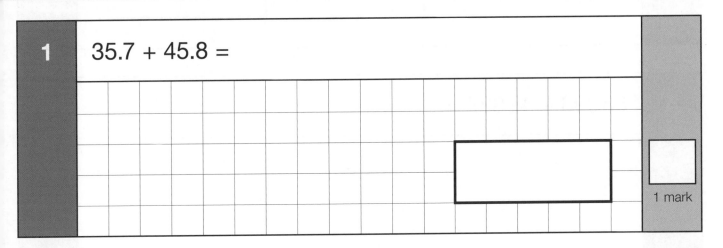

1 mark

2   807 + 100 =

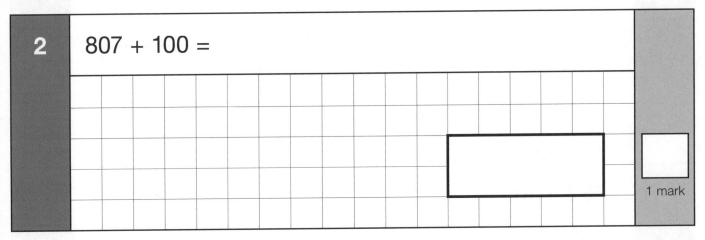

1 mark

3   5 × 6 × 8 =

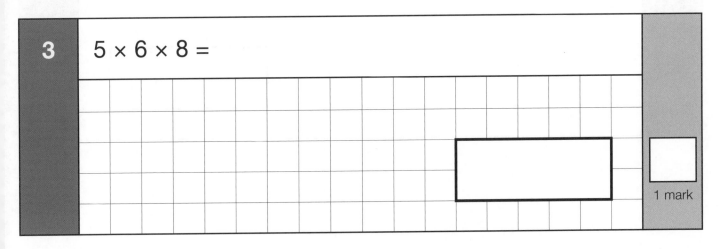

1 mark

4 160   170   180   190   _____

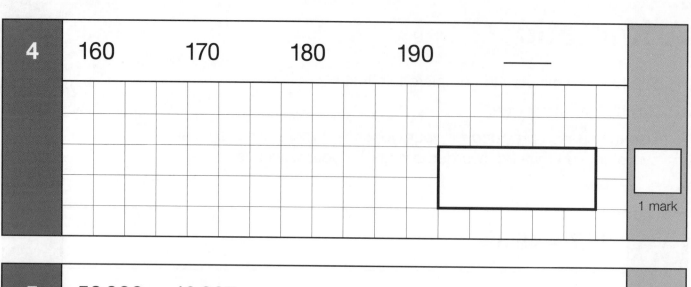

1 mark

5 53 093 + 42 827 =

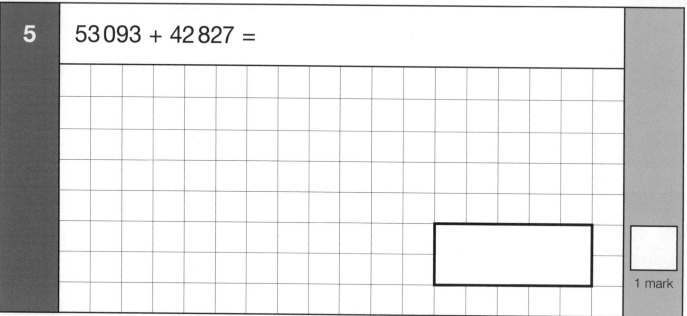

1 mark

6 57.32 × 100 =

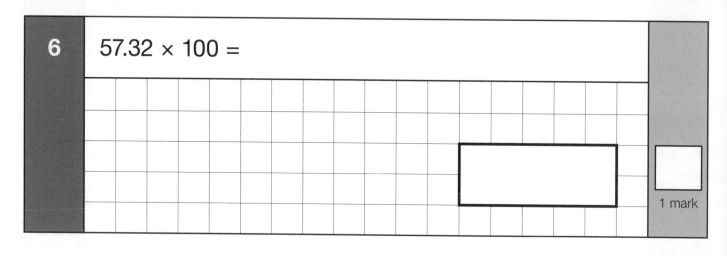

1 mark

7

$\dfrac{2}{5} \times 4 =$

1 mark

8

$71\% = \dfrac{\square}{\square}$

1 mark

9

$2825 \div 25 =$

1 mark

10

Show your working

```
    5 0 6
  ×   5 2
```

2 marks

11

$5\frac{7}{8} - 2\frac{6}{8} =$

1 mark

12

$\frac{8}{12} \div 4 =$

1 mark

13

Show your working

$$3\ 2\ \overline{|7\ 6\ 8}$$

2 marks

14

$(304 - 160) \div 24 + 24$

2 marks

Arithmetic

- You have 20 minutes to complete this test.
- Calculator <u>not</u> allowed.
- Use the spaces provided for your workings. Where two marks are available, you may be awarded a mark for your workings.

1 $\dfrac{11}{12} - \dfrac{\square}{\square} = \dfrac{4}{12}$

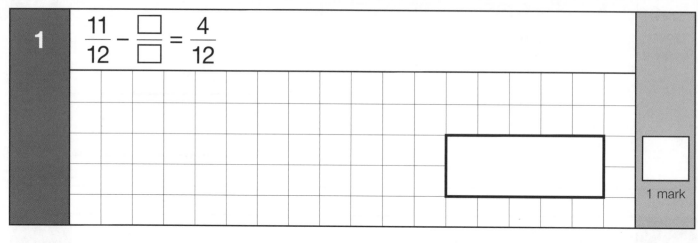

1 mark

2 $1386 + 257 - 406 =$

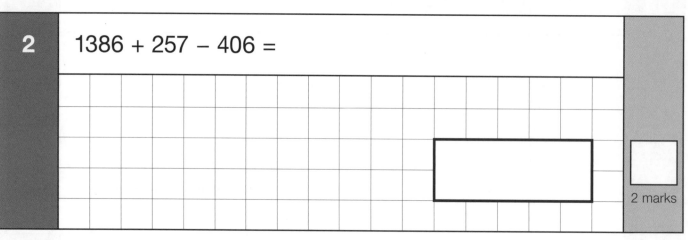

2 marks

3 $4^2 + 5^3 =$

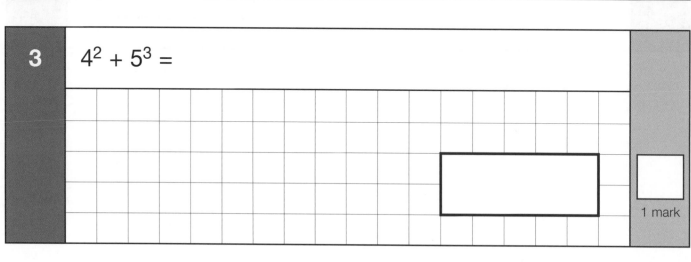

1 mark

4

$80 \times 16 = \boxed{} \times 10 + 80 \times \boxed{} = 1280$

1 mark

5

$81 + 75 = 243 - \boxed{}$

1 mark

6

$5076 \div 6 =$

1 mark

7 5000 ÷ 50 – 30 =

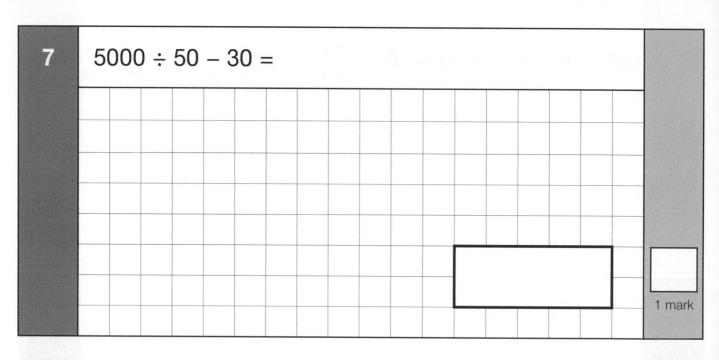

1 mark

8 $4\frac{3}{4} + 7\frac{1}{5} =$

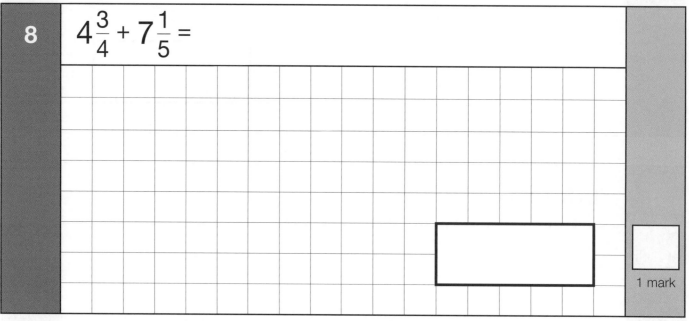

1 mark

9 $\frac{3}{8} \times \frac{2}{3} =$

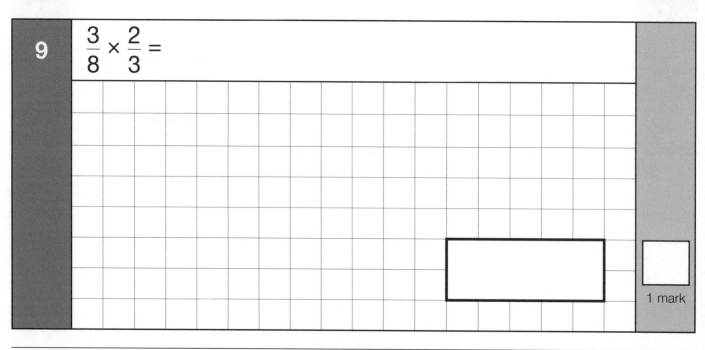

1 mark

10 0.09 × 4 =

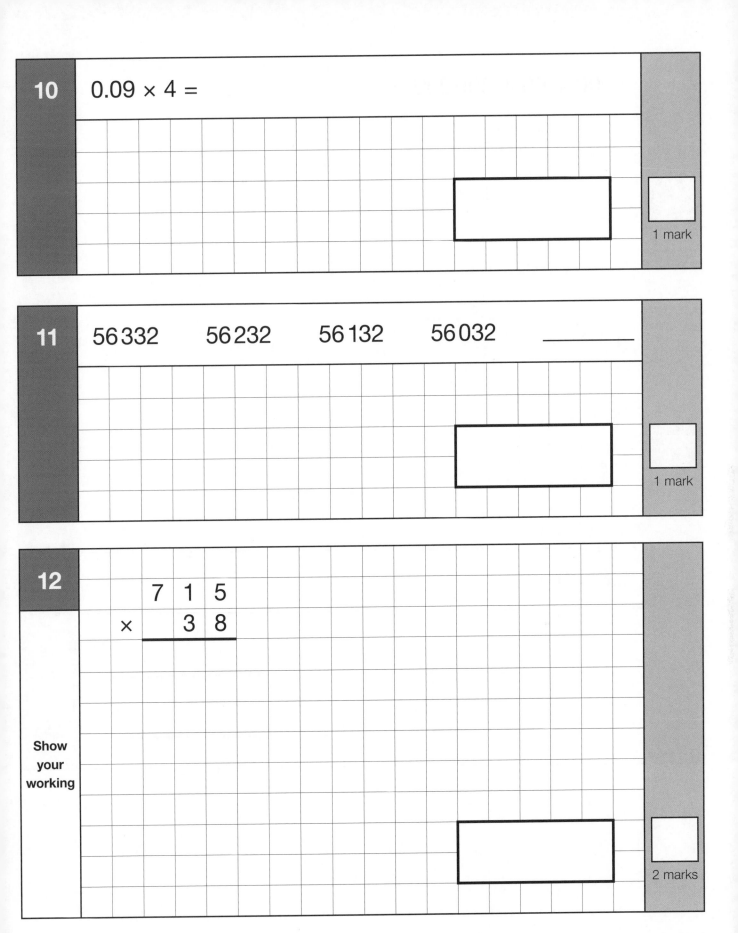

1 mark

11 56 332   56 232   56 132   56 032   _____

1 mark

12

Show your working

```
    7 1 5
×     3 8
```

2 marks

13   4000 + 60 + 300 000 =

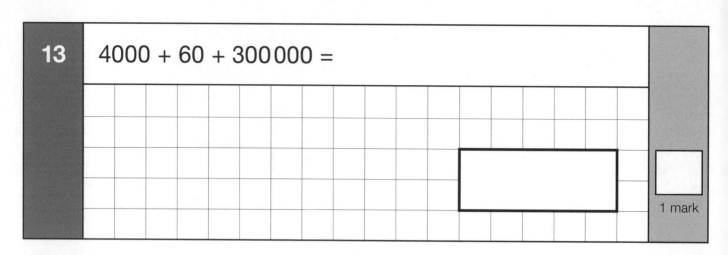

1 mark

14 $\dfrac{1}{8} \div 3 =$

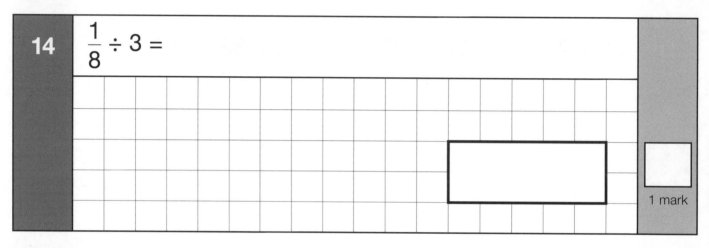

1 mark

15   0.8 × 9 =

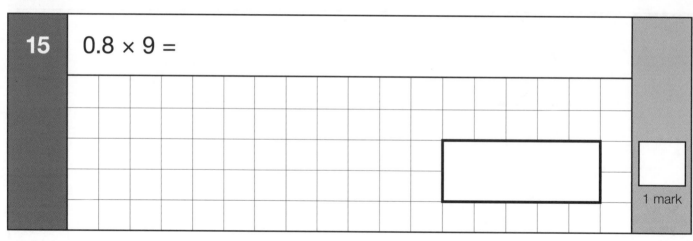

1 mark

Total _____ / 17 marks

Arithmetic

- You have 20 minutes to complete this test.
- Calculator not allowed.
- Use the spaces provided for your workings. Where two marks are available, you may be awarded a mark for your workings.

1 $736254 - 1000 =$

1 mark

2 $15 + \boxed{} + 65 = 115$

1 mark

3 $8760 + 2732 + 4689 =$

1 mark

4

$90 = 15 \times \boxed{}$

1 mark

5

$75 \times 10 = 0.75 \times \boxed{}$

1 mark

6

$\dfrac{2}{5} \times 6 =$

1 mark

7 $\dfrac{3}{4} = \boxed{}\ \%$

8 $57 \div 4 = \boxed{}\,\boxed{}\,.\,\boxed{}\,\boxed{}$

9 $0.7 \times 9 =$

10

$$\frac{1}{5} \div 2 =$$

1 mark

11

| 2 | 8 | 7 | 8 | 2 | 6 |

Show your working

2 marks

12

$$-7 + 17 =$$

1 mark

13

Show your working

$$
\begin{array}{r}
5\ 9\ 1\ 8 \\
\times \quad\quad 5\ 6 \\
\hline
\end{array}
$$

2 marks

14

$60.8 \div 100 =$

1 mark

15

$10\dfrac{1}{2} - 4\dfrac{4}{5} =$

1 mark

Total _____ / 17 marks

Arithmetic

- You have 20 minutes to complete this test.
- Calculator <u>not</u> allowed.
- Use the spaces provided for your workings. Where two marks are available, you may be awarded a mark for your workings.

1 $504\,008 = 8 + 4000 + \square$

1 mark

2 $5\dfrac{3}{8} = \dfrac{\square}{8}$

1 mark

3 $3 \times \dfrac{5}{6} =$

1 mark

4   518 + 684 + 3783 =

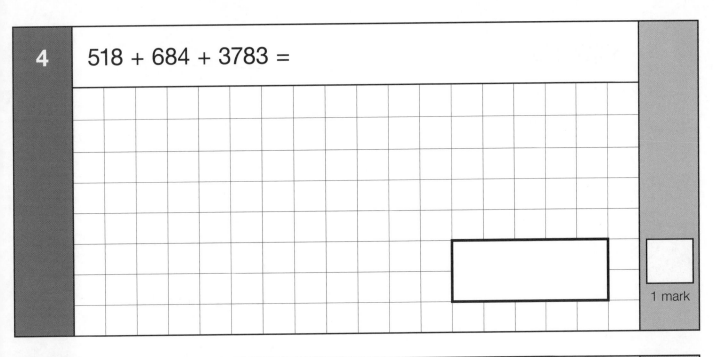

1 mark

5   5247 − 749 =

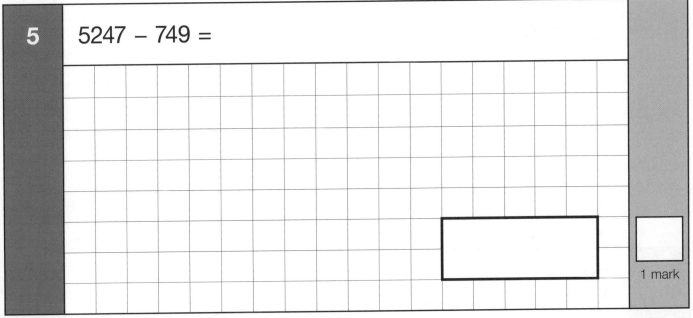

1 mark

6   400 000 = 4 × ▯

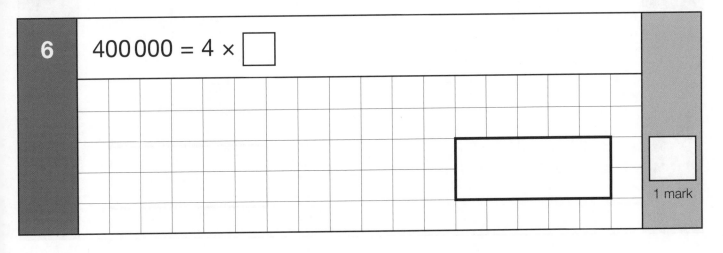

1 mark

7 $537.7 \div 5 =$

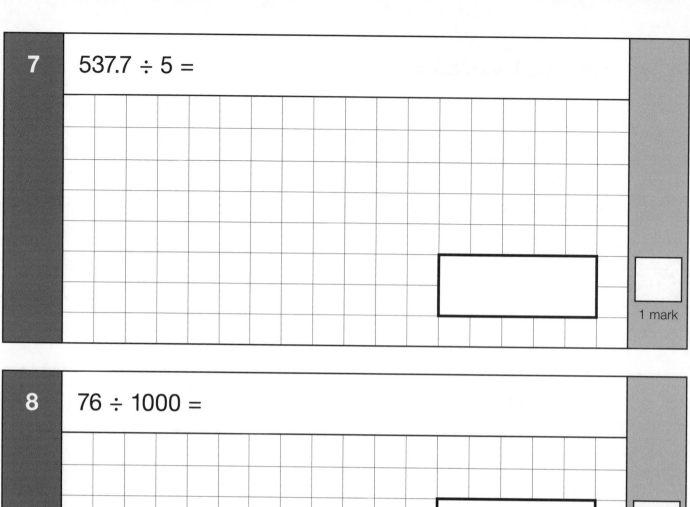

1 mark

8 $76 \div 1000 =$

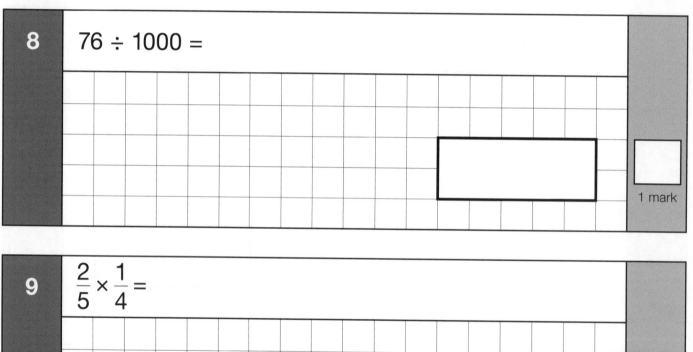

1 mark

9 $\dfrac{2}{5} \times \dfrac{1}{4} =$

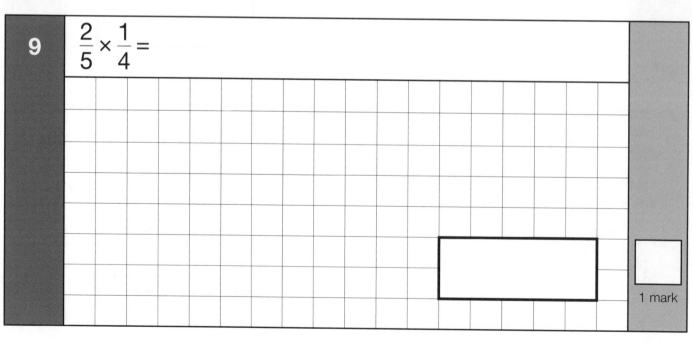

1 mark

10

$$6\frac{3}{10} + 5\frac{1}{4} =$$

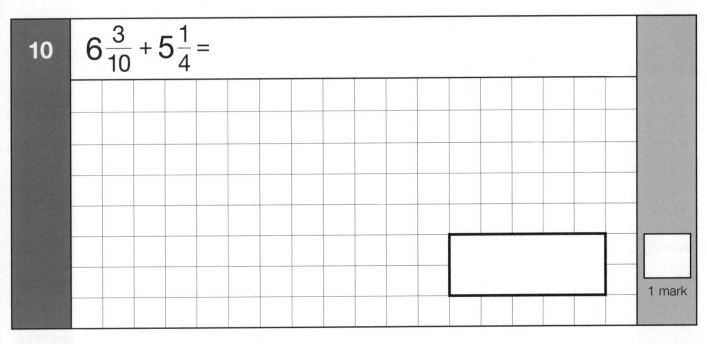

1 mark

11

$(17 + 14) \times (4 + 6 - 8) =$

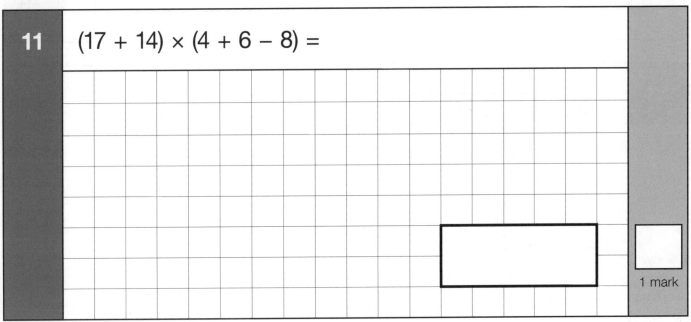

1 mark

12

Show your working

```
    7 5 9 3
  ×     7 4
```

2 marks

13

Show your working

```
2 4 ) 2 0 8 2
```

2 marks

Using Mathematics

- You have 20 minutes to complete this test.
- Calculator <u>not</u> allowed.

1 Which number comes next in this sequence?

48   56   64   72

1 mark

2

What time is shown on this clock?

1 mark

3 5727   5277   5272   5772

Write these numbers in order, smallest first.

smallest

1 mark

33

4 Part of this shape is missing.

The dotted line is a line of symmetry.

> Complete the shape.

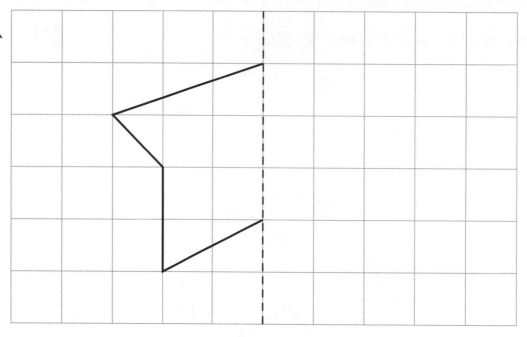

5 Tom has 217 minutes left on his phone.

He uses 83 minutes.

He gets another 350 minutes.

> How many minutes does Tom have on his phone?

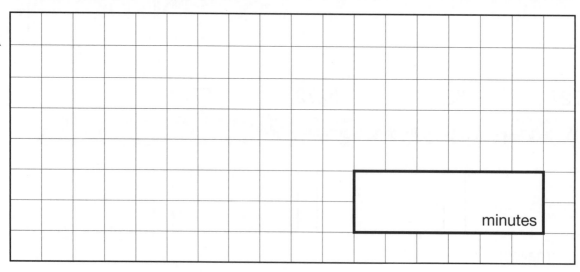

minutes

6 Four hundred and sixty thousand, three hundred and five.

Write this number in digits.

1 mark

7 Two prime numbers total 31

What are the two numbers?

1 mark

8 Write the equivalent fractions shown by the shading in these shapes.

$$\frac{\square}{\square} = \frac{\square}{\square}$$

2 marks

9 This is a train timetable.

Little Oak	10.30	11.30	12.30	13.30
Elmstree		12.10		
Ashton	11.25		13.25	14.25
Beechwood			13.43	14.41
Birchly	12.03	12.58	14.05	15.06

Jack arrives at Little Oak at 11.00

He catches the next train to Ashton.

When will he arrive at Ashton?

1 mark

10 What is 25% of 600?

1 mark

11 Aisha adds two numbers; the total is 15

The difference between the two numbers is 5

What are the two numbers?

1 mark

12 This is a cube.

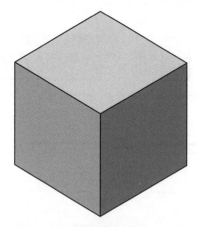

Complete these sentences.

A cube has [ ] faces.

A cube has [ ] edges.

A cube has [ ] vertices.

3 marks

Total _____ / 15 marks

Using Mathematics

- You have 20 minutes to complete this test.
- Calculator not allowed.

---

1 Which three-digit number can you make from these three cards?

3	4	9
units	hundreds	tens

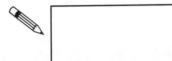

1 mark

---

2 Tick (✔) the right angles in this shape.

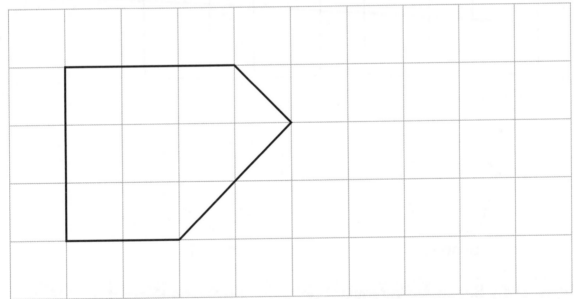

2 marks

---

3 Round 5199 to the nearest thousand.

1 mark

---

4 Ben sees MCMLXXX written on a building.

MCMLXXX stands for a year.

Name the year.

1 mark

5 Circle the factors of 30.

3   4   6   15   60   150

1 mark

6 Write these fractions in order, starting with the largest.

$$\frac{3}{5} \qquad \frac{7}{10} \qquad \frac{61}{100} \qquad \frac{13}{20}$$

largest

1 mark

7

Shape A   Shape B

Explain why Shape A is a regular pentagon and Shape B is not.

_____

_____

_____

1 mark

8   1 inch is about 2.5 centimetres.

How many centimetres is 12 inches?

cm

1 mark

9   The temperature outside a greenhouse is –4°C.

The temperature inside the greenhouse is 4°C.

What is the difference between the two temperatures?

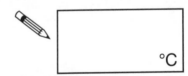

°C

1 mark

10 William sees a special offer in a shop.

He can buy four cans of soup for £1.50

How much would William pay for twelve cans of soup?

£

1 mark

11 Here are three number cards:

| 7 | 1 | 6 |

Nishi arranges the cards to make three-digit numbers.

List all the odd numbers that Nishi can make.

2 marks

12 Find the missing number.

$0.7 \times \boxed{} = 4.2$

1 mark

13 This pie chart shows the sports chosen by 80 children.

Here are some facts about the pie chart:

- 30 children chose tennis.

- $\frac{1}{4}$ of the children chose football.

- The same number of children chose gymnastics as chose rugby.

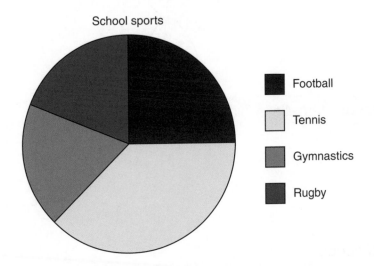

School sports

Football

Tennis

Gymnastics

Rugby

How many children chose rugby?

Show your working

children

2 marks

Total _____ / 16 marks

Using Mathematics

- You have 20 minutes to complete this test.
- Calculator not allowed.

1 This bag has 5 black balls and 3 white balls.

What fraction of the balls is black?

1 mark

2 A train has 8 coaches.

Each coach has 72 seats.

The ticket collector says, 'I know 70 × 8 = 560'

What must he add to 560 to find how many seats there are in the train altogether?

1 mark

3 **a.** Round 5.17 to:
i. one decimal place.   **ii.** the nearest whole number.

1 mark

b. Round 14.73 to:
i. one decimal place.   **ii.** the nearest whole number.

1 mark

Draw lines to match the same lengths.

One has been drawn for you.

4.6m	4.6cm
46mm	0.46km
460m	460cm
4.6mm —————————	0.46cm
46cm	0.46m

2 marks

5 Polly calculates 87 × 73

Round each number to the nearest ten.

Use the rounded numbers to give an estimated answer to Polly's calculation.

1 mark

6 What number comes next in this sequence?

536 280   537 280   538 280   539 280

1 mark

7

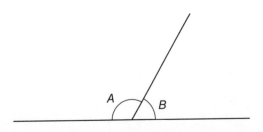

What is the total of angle A and angle B?

o

1 mark

8 In 2 hours a machine makes 800 packs of biscuits.

64 packs of biscuits fill a box.

a. How many boxes will be filled completely in 2 hours?

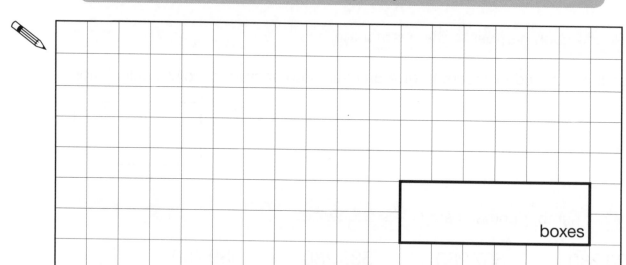

boxes

1 mark

b. How many packs of biscuits will be made in 12 hours?

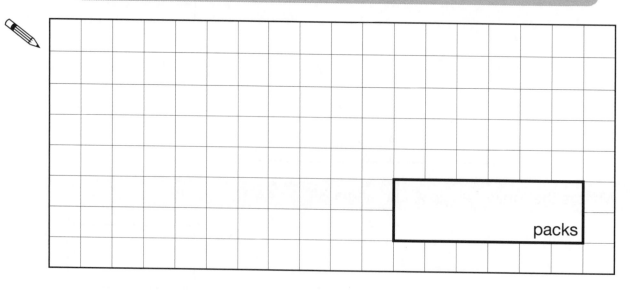

packs

1 mark

9 Josef says, 'I am thinking of a number. I multiply the number by 6 and subtract 8'

Tick (✔) the expression that shows Josef's calculation algebraically.

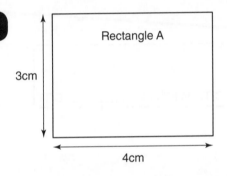

☐ 6n + 8

☐ 8 – 6n

☐ 6(n – 8)

☐ 6n – 8

☐ 8 × 6 – n

1 mark

10

Rectangle A

3cm

4cm

Rectangle B

x

8cm

Rectangle A is enlarged and is drawn as rectangle B.

How long is side x?

cm

1 mark

11

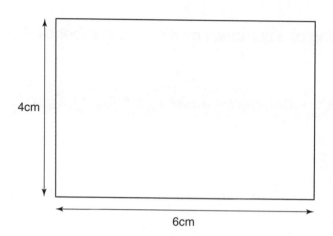

4cm

6cm

a. Give the length and width of a different rectangle with the same area.

length = _____ cm, width = _____ cm

1 mark

b. Give the length and width of a different rectangle with the same perimeter.

length = _____ cm, width = _____ cm

1 mark

12 Circle the fractions that will simply to $\frac{3}{5}$

$\frac{45}{75}$ $\frac{18}{30}$ $\frac{10}{15}$ $\frac{25}{40}$ $\frac{36}{60}$

2 marks

Total _____ / 17 marks

Using Mathematics

- You have 20 minutes to complete this test.
- Calculator <u>not</u> allowed.

1 This is a rectangle.

Tick (✔) **two** correct statements.

☐ The two bold lines are perpendicular.

☐ The bold and thin lines are perpendicular.

☐ The two bold lines are parallel.

☐ The bold and thin lines are parallel.

2 marks

2 Put these numbers in order, starting with the smallest.

39.39	9.93	3.09	9.09	3.39

smallest

1 mark

3 Circle the **square** numbers.

1   5   36   46   64   91

4 This table shows Obi's journey from York to Carlisle.

	Time taken (minutes)
York to Newcastle	55
Waiting time	35
Newcastle to Carlisle	50

How long did it take Obi to go from York to Carlisle in hours and minutes?

Show your working

___ hours ___ minutes

5 The perimeter of this rectangle is 56cm.

The length of the rectangle is 17cm.

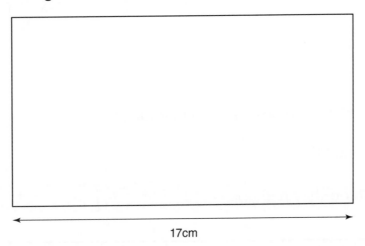

17cm

Calculate the width of the rectangle.

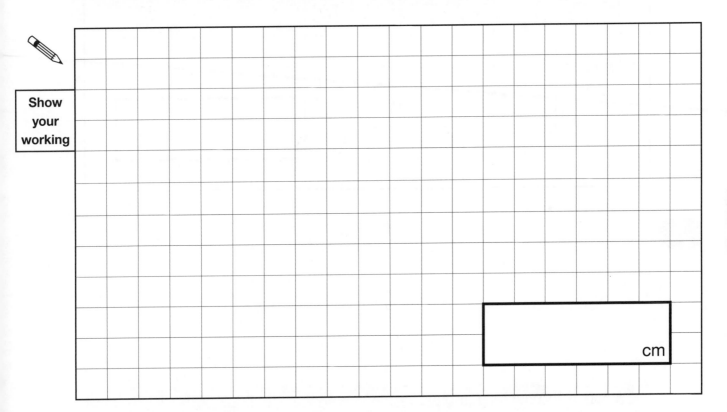

Show your working

cm

2 marks

6 Simplify $\frac{16}{24}$

7 Four friends share the cost of a meal equally.

The meal costs £63

How much do they each pay?

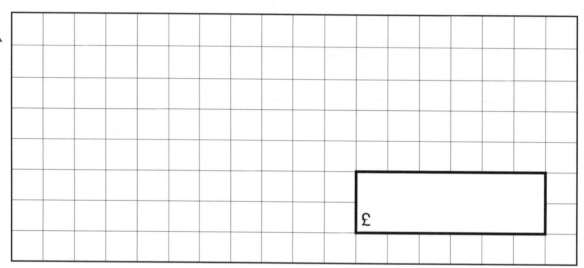

£

8

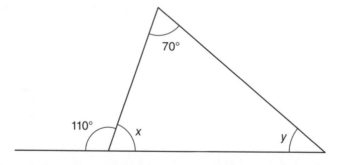

70°

110° x

y

Calculate the missing angles.

Angle x = 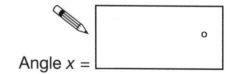 °

Angle y =  °

9 A car hire firm charges its customers, in pounds (£), by the day and for every mile driven.

They use the formula 10*d* + 0.1*m*

where *d* = the number of days and *m* = the number of miles

How much would the car hire firm charge for renting a car for 3 days and driving 50 miles?

Show your working

£

2 marks

Total _____ / 15 marks

Using Mathematics

- You have 20 minutes to complete this test.
- Calculator <u>not</u> allowed.

1 This bar chart shows the number of minutes Mohammed spends on his mobile phone in one week.

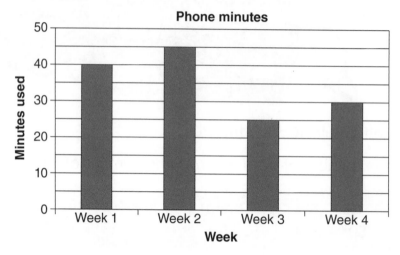

Mohammed's mobile phone contract allows him 200 minutes for four weeks.

How many minutes does Mohammed have left?

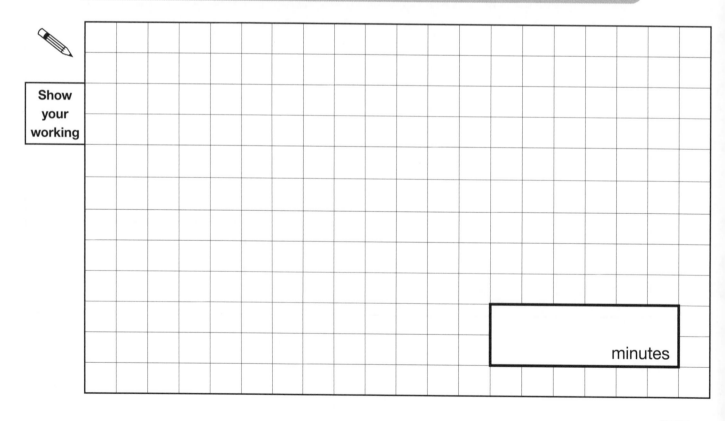

Show your working

minutes

2 marks

2 Sonny uses two £20 notes to buy four books that cost £6.99 each.

How much change does he get?

£ [ ]

3 Tick (✔) the number which has a 3 with the value of thirty thousand.

[ ] **3**47 268

[ ] 876 0**3**8

[ ] 9**3**1 246

[ ] 743 971

[ ] 275 384

4 Dev sat three tests.

His results were:

Maths $\frac{17}{20}$

English $\frac{4}{5}$

Science $\frac{42}{50}$

Dev uses percentages to compare his results.

What was his best result as a percentage?

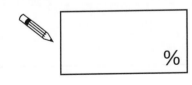

% [ ]

5 Once a month, a lorry delivers a load of goods to a shop.

The whole journey is 376 miles.

a. How many miles does the lorry travel over 12 months?

miles

1 mark

b. The lorry uses a gallon of fuel every 8 miles.

How many gallons does the lorry use on **each** journey?

gallons

1 mark

6 Max makes some concrete for a path.

For a path 8 metres long Max needs:

- 200kg of cement

- 600kg of sand

- 600kg of stone

What weight of stone will he need for a path 20 metres long?

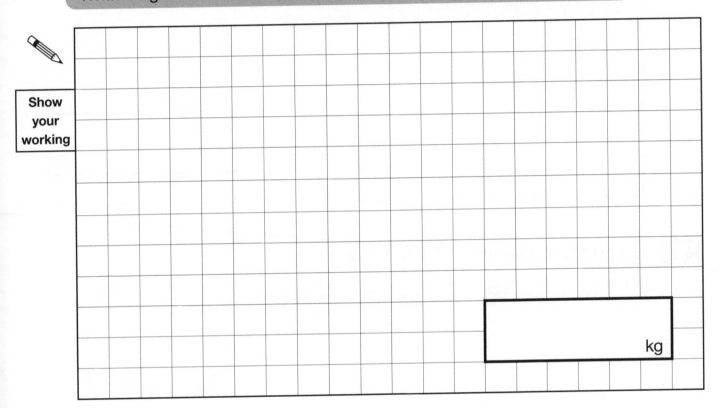

Show your working

kg

2 marks

7 Write the missing numbers:

$$\frac{1}{2} \times \frac{1}{\boxed{}} = \frac{1}{4} = \frac{1}{\boxed{}} \div 2$$

2 marks

8 Manisha makes a drink using orange juice and lemonade.

She uses three times as much lemonade as orange juice.

She makes 600ml of drink.

How much lemonade will she need?

ml

9 Approximately, 1 kilometre equals 0.6 miles.

Use this approximation to change 80 kilometres into miles.

miles

10 Complete the labels to name the parts of a circle.

The centre is marked by a black dot.

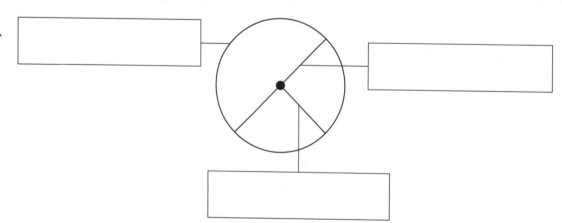

Total _____ / 15 marks

Using Mathematics

- You have 20 minutes to complete this test.
- Calculator not allowed.

1 This is part of a calendar.

June						
Sun	**Mon**	**Tue**	**Wed**	**Thu**	**Fri**	**Sat**
			1	2	3	4
5	6	7	8	9	10	11

Emily's birthday was on the last Saturday in May.

What was the date of Emily's birthday?

1 mark

2 Carla looks at a train timetable.

Her train leaves at 16.08

Carla arrives at the station at ten to four.

How long is it before her train leaves?

minutes

1 mark

3 The population of a city was 97 358

One hundred years later, the population was 213 064

a. By how many had the population increased?

b. Another ten years later, the population had increased by another 37 786

What was the population ten years later?

4 Put these numbers in order, starting with the largest.

2.903   2.593   2.33   2.9

largest

5 **a.** Calculate the missing number.

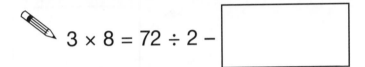

$56 \times 9 = 1512 \div$ ⬚

1 mark

b. Calculate the missing number.

$3 \times 8 = 72 \div 2 -$ ⬚

1 mark

6 Calculate the perimeter of this shape.

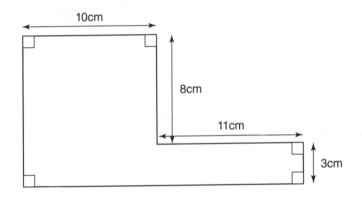

10cm

8cm

11cm

3cm

cm

1 mark

7 Calculate angle *a*.

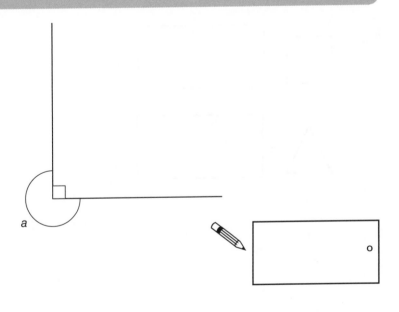

a

°

1 mark

8 **a.** Round 319 452 to the nearest hundred.

1 mark

b. Round 319 452 to the nearest hundred thousand.

1 mark

9 Tick (✔) the number that is a common multiple of 12 and 15

| 3 | 5 | 30 | 60 | 75 |

1 mark

10 These shapes stand for numbers.

☐ × △ = 36

△ ÷ ☐ = 4

Which numbers do the shapes stand for?

☐ =

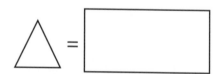

△ =

1 mark

11 Two vertices of a square are marked on the grid with the symbol ●

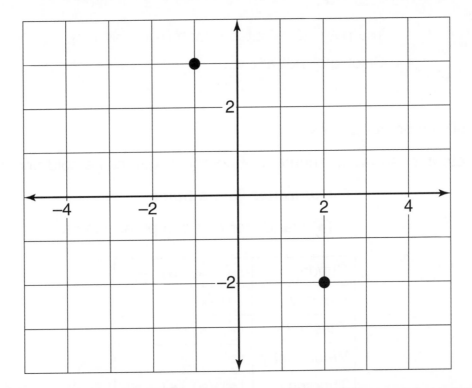

a. Mark a third vertex at (3, 2).

1 mark

b. What are the coordinates of the fourth vertex?

(____ , ____)

1 mark

12 These fractions are in order of size, smallest first.

Write the missing numbers.

$\dfrac{7}{12}$ $\dfrac{\square}{24}$ $\dfrac{2}{3}$ $\dfrac{\square}{6}$ $\dfrac{7}{8}$

2 marks

Total _____ / 17 marks

Using Mathematics

- You have 20 minutes to complete this test.
- Calculator <u>not</u> allowed.

1 Dom sells computer games.

This pictogram shows the number of computer games he sold one week.

Computer Game Sales

⊙ stands for four computer games

Sunday	⊙ ⊙ ⊙ ⊙
Monday	⊙ ◖
Tuesday	⊙ ◖
Wednesday	⊙
Thursday	⊙ ⊙ ⊙
Friday	⊙ ⊙ ⊙ ◟
Saturday	⊙ ⊙ ⊙ ⊙

a. How many computer games did Dom sell on Friday and Saturday?

games

1 mark

b. How many more games did Dom sell on Sunday than on Monday?

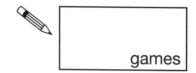

games

1 mark

2 Circle the prime numbers.

  19   29   39   49   59   69

2 marks

3

What time is shown on the clock?

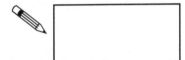

1 mark

4

$$\frac{1}{2} + \frac{11}{12} + \frac{3}{4} =$$

Give your answer as a mixed number with simplified fraction.

Show your working

3 marks

5

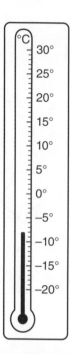

What will the temperature be after the temperature shown on the thermometer has gone up by 3°C?

°C

1 mark

6 Here is part of a number sequence.

The numbers decrease in equal steps.

Fill in the missing numbers.

 58 [ ] [ ] 40 [ ] 28

2 marks

7 Put these fractions in order, starting with the smallest.

$\frac{3}{4}$ $\frac{5}{6}$ $\frac{1}{2}$ $\frac{7}{12}$ $\frac{2}{3}$

 [ ] [ ] [ ] [ ] [ ]

smallest

1 mark

8 Write six million, two hundred thousand, three hundred and fifteen in figures.

9 This shape is a parallelogram.

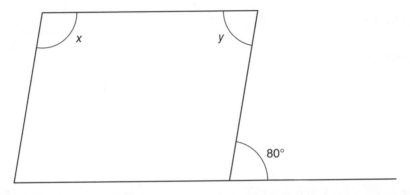

Calculate angle *x* and angle *y*.

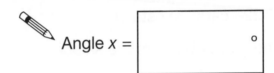

Angle *x* = ◻ °

Angle *y* = ◻ °

10 Calculate the missing numbers.

 a. 15% of 250 = ◻

b. 15% of ◻ = 90

Total _____ / 17 marks

Using Mathematics

- You have 20 minutes to complete this test.
- Calculator <u>not</u> allowed.

1 Ellie has a 2 litre container of juice.

She pours equal amounts of juice into 8 glasses.

How much juice is in each glass?

Give your answer in litres.

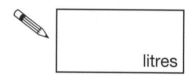

litres

1 mark

2 A car firm builds 86 258 cars a year.

35 453 cars are hatchbacks.

27 328 cars are estates.

The rest of the cars are saloons.

How many saloons did the car firm build?

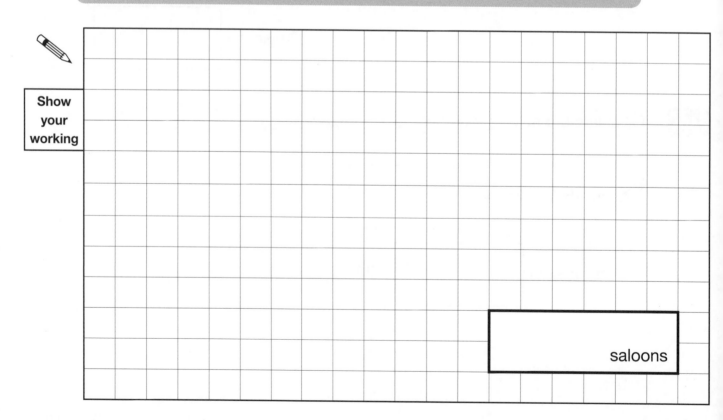

Show your working

saloons

2 marks

3 A factory can print 1632 books every 6 hours.

a. How many books will be printed in 1 hour?

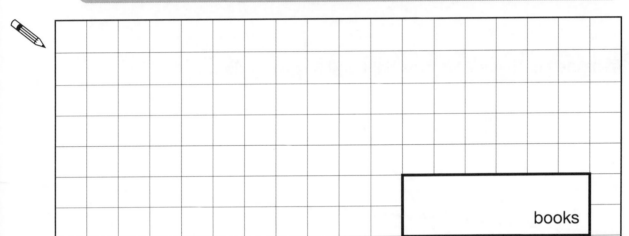

books

1 mark

b. How many books will be printed in 24 hours?

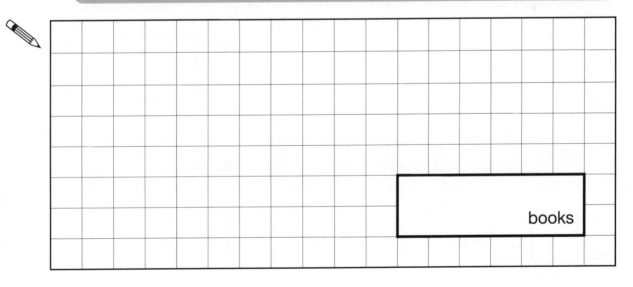

books

1 mark

4

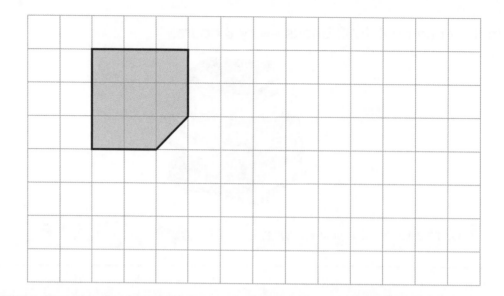

Translate the shape 7 squares right and 3 squares down.

2 marks

5

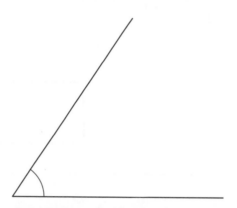

Measure the angle.

 °

1 mark

6

Tick (✔) the correct calculation to change $\frac{2}{5}$ into a decimal.

☐ 5 ÷ 2 =   ☐ 5 × 2 =

☐ 5 − 2 =   ☐ 2 ÷ 5 =

☐ 2 + 5 =

1 mark

7 Ned has a 10kg bag of potatoes.

He uses 2.3kg of potatoes one day.

He uses 1600g of the potatoes on the next day.

What is the weight of potatoes left?

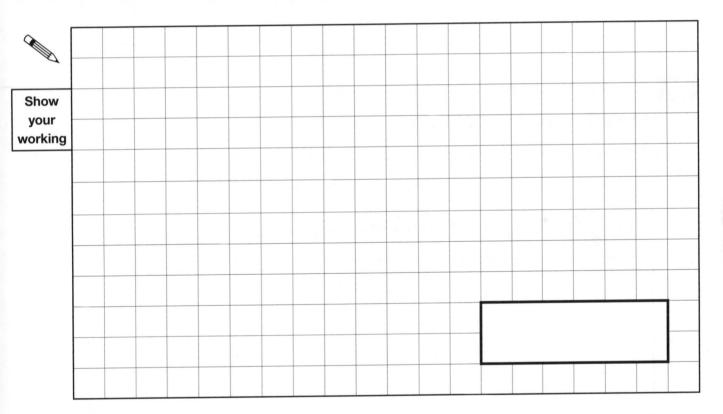

8 Write the missing number.

$$\frac{1}{3} \div \boxed{} = \frac{1}{12}$$

9 Name the 3-D shape this net will make.

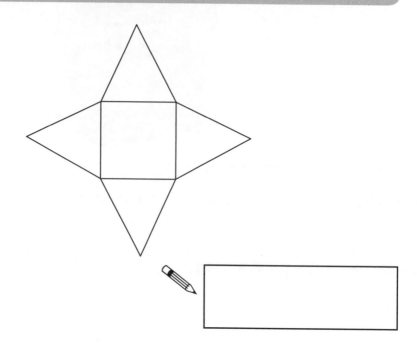

1 mark

10 A factory has 2 machines that bottle drinks.

One machine can fill 2365 bottles in one hour.

A newer machine can fill 3850 bottles in one hour.

The bottles are put into boxes holding 24 bottles.

How many boxes are completely filled in one hour?

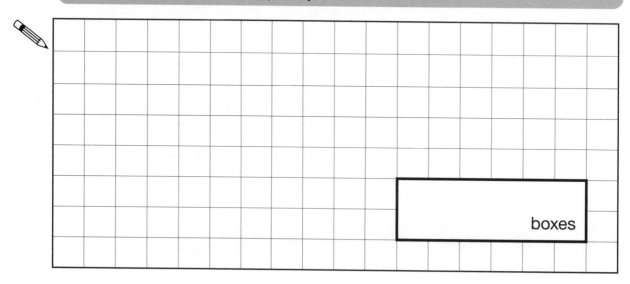

boxes

2 marks

Total _____ / 15 marks

Using Mathematics

- You have 20 minutes to complete this test.
- Calculator not allowed.

1 48 boxes can fit into 5 vans.

How many boxes can fit into 20 vans?

| boxes
1 mark

2 Complete the missing numbers.

78 ÷ 6 = (⬜ ÷ 6) + (18 ÷ ⬜) = 13

1 mark

3 Draw lines to join a decimal with its equivalent fraction.

1.3 $\dfrac{13}{100}$

0.103 $\dfrac{13}{10}$

0.13 $\dfrac{103}{1000}$

0.013 $\dfrac{103}{100}$

1.03 $\dfrac{13}{1000}$

2 marks

4 Outside temperatures were recorded at midday for a week.

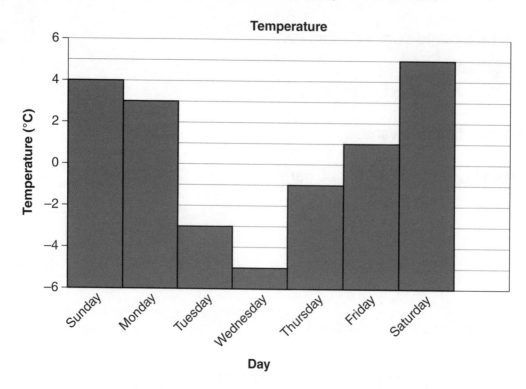

a. What was the temperature at midday on Monday?

°C

1 mark

b. What was the difference between the temperatures recorded on Sunday and Wednesday?

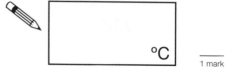

°C

1 mark

c. For how many days was the temperature below 0°C?

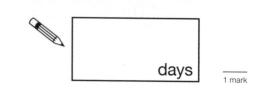

days

1 mark

5

$$7\,569\,135$$

Milly makes some changes to this number.

She writes:

- 4 as the tens of thousands digit

- 2 as the thousands digit

- 8 as the millions digit

She leaves the other numbers as they are.

Write Milly's new number.

6 What is the difference between the area of the square and the area of the oblong?

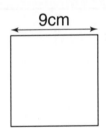 9cm

12cm

6cm

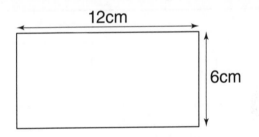 cm²

7 Calculate the missing number.

  ÷ 26 = 529

8 Calculate

 a. 20% of 600 = ☐

b. 90% of ☐ = 270

c. ☐ % of 200 = 50

9 Jo sits 6 maths tests.

Her marks were:

14   16   14   17   19   16

What was her mean mark?

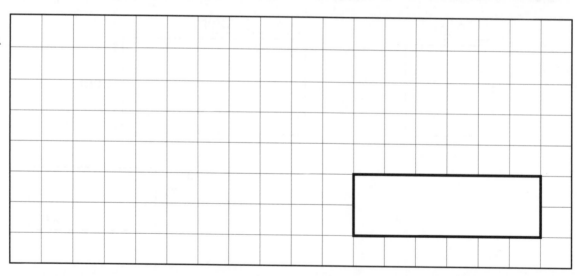

Total _____ / 15 marks

Using Mathematics

- You have 20 minutes to complete this test.
- Calculator <u>not</u> allowed.

---

1 Here are some number cards.

| 0 | 5 | 5 | 6 | 9 |

Rearrange the number cards to make the closest number to 60 000

1 mark

2 Write the missing number.

 $5.156 +$ ⬜ $= 12.48$

1 mark

3 Write the missing numbers.

 $70\% = \dfrac{7}{⬜}$

$⬜\% = \dfrac{3}{4}$

$80\% = \dfrac{⬜}{5}$

2 marks

---

4 The dotted line is the mirror line.

Reflect the shape in the mirror line.

Use a ruler.

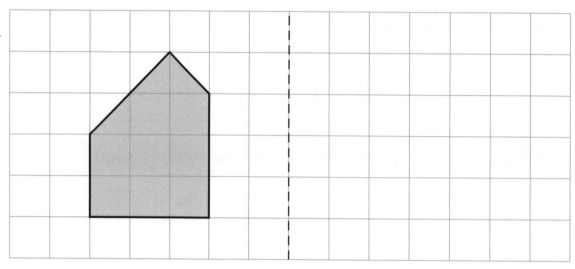

2 marks

5 What could the missing numbers be?

 $\dfrac{1}{\boxed{}} \times \dfrac{1}{\boxed{}} = \dfrac{1}{12}$

1 mark

6 Find the missing numbers.

 a. $3.2 \div 100 = \boxed{} \div 1000$

1 mark

b. $4.7 \times 10 = \boxed{} \times 1000$

1 mark

7 Debbie runs a cake shop.

She works out the cost of a cake, in pounds (£), by doubling the cost of the ingredients, *i*, and adding 4.

Tick (✔) the formula Debbie uses.

☐ 2 × 4 + *i*

☐ 2(4 + *i*)

☐ 2*i* + 4

☐ 4 + *i* + 2

☐ 4*i* + 2

1 mark

8 A rectangle is 12cm long and 5cm wide.

The rectangle is enlarged by a scale factor of 4

What are the length and width of the enlarged rectangle?

length = _____ cm, width = _____ cm

1 mark

9

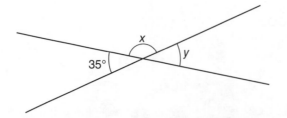

Calculate angle *x* and angle *y*.

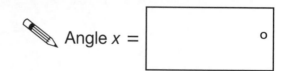

Angle *x* = ☐ °

Angle *y* = ☐ °

10

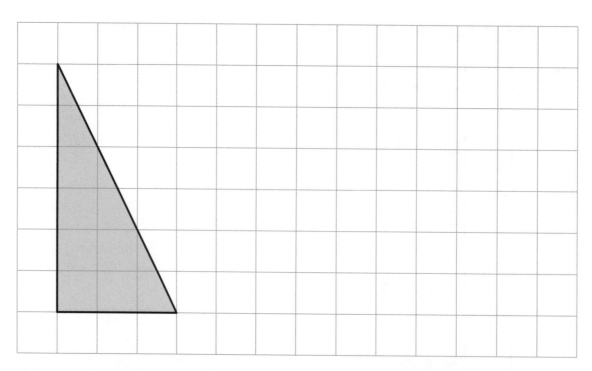

a. Calculate the area of the triangle.

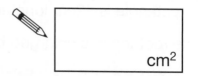

☐ cm²

b. Draw a **square** on the grid with the same area as the triangle.

Use a ruler.

Total _____ / 15 marks

Using Mathematics

- You have 20 minutes to complete this test.
- Calculator not allowed.

1 Ola adds two fractions.

The answer is $1\frac{1}{4}$

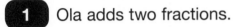 Tick (✔) the two fractions Ola adds.

☐ $\frac{1}{4} + \frac{1}{4}$

☐ $\frac{4}{5} + \frac{2}{5}$

☐ $\frac{9}{12} + \frac{7}{12}$

☐ $\frac{3}{8} + \frac{7}{8}$

☐ $\frac{7}{20} + \frac{17}{20}$

1 mark

2 Circle the number with a 6 as hundreds of thousands **and** a 4 as thousands.

 543 619   6 456 752   2 634 891   645 894   1 mark

3 **a.** 7.4 has been rounded to one decimal place.

Tick (✔) the number that was rounded.

7.32	7.35	7.45	7.49
☐	☐	☐	☐

1 mark

b. 10 has been rounded to the nearest whole number.

Tick (✔) the number that was rounded.

9.19	9.48	10.09	10.55
☐	☐	☐	☐

1 mark

4 Calculate the missing numbers.

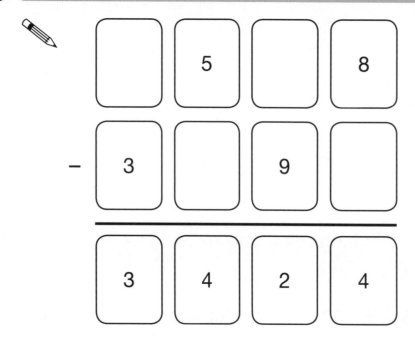

1 mark

5 This is a menu for a takeaway.

Meal	Price
Pizza – 12 inch	£9.50
Pizza – 10 inch	£7.00
Fish	£3.20
Large chips	£1.50
Small chips	90p

Judy uses a £20 note to buy:

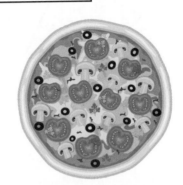

- a 10 inch pizza
- 2 fish
- 1 large chips
- 1 small chips

How much change will Judy get?

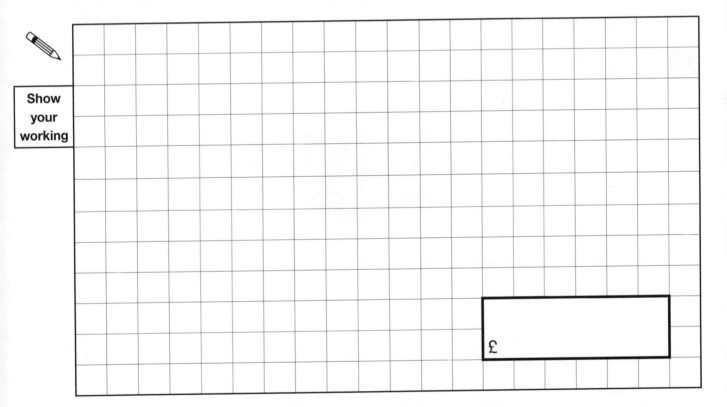

Show your working

£

2 marks

6 The two dotted lines are both mirror lines.

A shape has been reflected in both lines.

> Draw the two missing shapes from the grid.

Use a ruler.

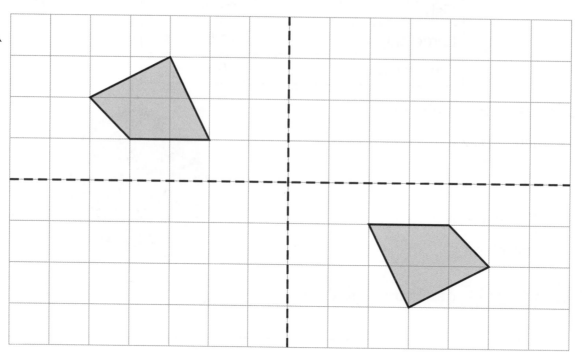

2 marks

7 Fay can buy a 2kg bag of 14 apples for £3

a. How many apples can Fay buy for £1.50?

apples

1 mark

b. How many apples would Fay have if she bought 5kg?

apples

1 mark

8 Here are two fair triangular spinners.

Each spinner is spun once and the numbers added to give a total.

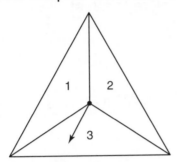
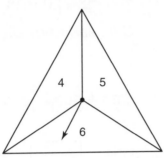

How many different totals can be made?

9 Find the missing numbers.

$$0.06 \times \boxed{} = 0.36 = \frac{36}{100} = \frac{\boxed{}}{25}$$

10 The triangle is an **equilateral** triangle.

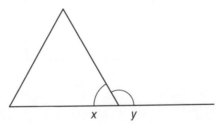

Calculate angle x and angle y.

Angle $x = \boxed{}$ °

Angle $y = \boxed{}$ °

Total _____ / 16 marks

Using Mathematics

- You have 20 minutes to complete this test.
- Calculator not allowed.

---

1 Complete the missing numbers from the equivalent fractions.

a. $\dfrac{3}{4} = \dfrac{6}{\boxed{}} = \dfrac{\boxed{}}{12} = \dfrac{12}{\boxed{}}$

1 mark

b. $\dfrac{5}{8} = \dfrac{\boxed{}}{16} = \dfrac{15}{\boxed{}} = \dfrac{\boxed{}}{32}$

1 mark

2 A number squared and a number cubed both equal 64.

Find the numbers.

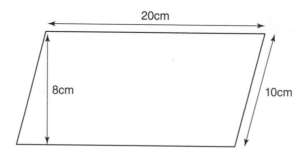

$\boxed{}^2 = 64 = \boxed{}^3$

1 mark

3 Calculate the area of this parallelogram.

20cm

8cm

10cm

cm^2

1 mark

---

4 Sam sells t-shirts from a market stall.

Each t-shirt costs £8

Sam goes to 12 markets.

The mean number of t-shirts sold at the markets is 20

How much did Sam sell all the t-shirts for?

Show your working

£

2 marks

5 This is a map of an island.

It is drawn on a centimetre square grid.

Each square centimetre represents 1 square kilometre.

What is the approximate area of the island?

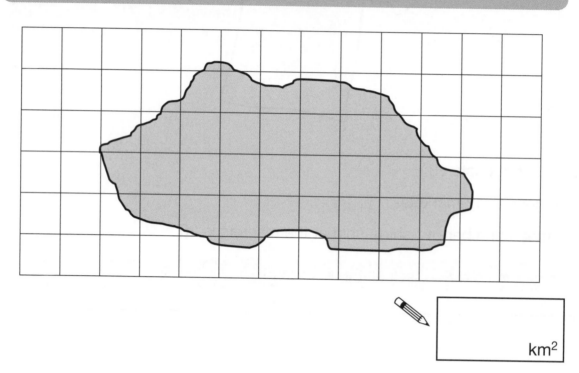

km²

6 This table gives approximate conversions between kilograms and pounds.

kilograms	pounds
1	2.2
2	4.4
4	8.8
8	17.6
16	35.2

Use the approximations to convert 7 kilograms into pounds.

pounds

7 Calculate the volume of this cuboid.

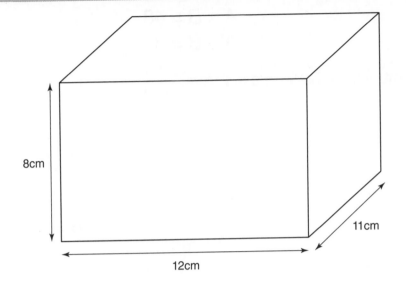

8cm

12cm

11cm

 cm³

8 Calculate the missing numbers.

 a. $4\frac{1}{2} +$ ☐ $= 10\frac{7}{8}$

b. $8\frac{11}{12} -$ ☐ $= 5\frac{1}{12}$

9 This sequence decreases in equal steps.

Find the missing number.

 6   −3   ☐   −17   −26

10 The letters stand for numbers.

$$A \times B = 48$$
$$A \div B = 3$$

Which numbers do the letters stand for?

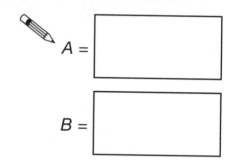

$A =$

$B =$

2 marks

11 The points of 3 vertices of a parallelogram are marked on the grid with the symbol ●

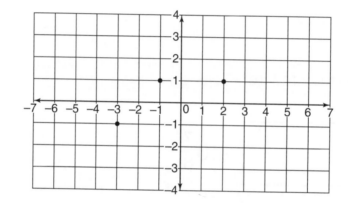

a. Give the coordinates of the 3 marked vertices.

(,   ) (,   ) (,   )

1 mark

Draw lines to complete the parallelogram.

Use a ruler.

b. What could the coordinates of the fourth vertex be?

(,   )

1 mark

Total _____ / 16 marks

Arithmetic: Answers

Test 1

1. 579 **(1 mark)**
2. 8103 **(1 mark)**
3. $1\frac{6}{10}$ (Accept $\frac{16}{10}$) **(1 mark)**
4. 604 **(1 mark)**
5. 84 **(1 mark)**
6. 18444 **(1 mark)**
7. $\frac{13}{5}$ **(1 mark)**
8. $\frac{5}{12}$ **(1 mark)**
9. 50 **(1 mark)**
10.

		3	2	8
	×		2	6
1	9	6	8	
6	5	6	0	
8	5	2	8	

(2 marks: 1 mark for using long multiplication with no more than one error; 1 mark for correct answer. Do not award any marks if the 0 for multiplying by a ten is missing.)

11.

				7	1
2	5	1	7	7	5
		1	7	5	
				2	5
				2	5
					0

(2 marks: 1 mark for using long division with no more than one error; 1 mark for correct answer)

12. 102 **(1 mark)**
13. −8 **(1 mark)**
14. $\frac{1}{15}$ **(1 mark)**
15. 0.49 **(1 mark)**

Test 2

1. 597 **(1 mark)**
2. 2 **(1 mark)**

3. 3888 **(1 mark)**
4. 1369 **(1 mark)**
5. 1000 **(1 mark)**
6. 2490 **(1 mark)**
7. 3702 **(1 mark)**
8. 670 **(1 mark)**
9. $3\frac{3}{4}$ (Accept $\frac{15}{4}$) **(1 mark)**
10. −17 **(1 mark)**
11. 4480

(2 marks: 1 mark for reaching 560 × 8 (calculation in brackets, 305 − 297, must be completed first); 1 mark for correct answer)

12. 3 **(1 mark)**
13.

		6	4	8
	×		3	5
	3	2	4	0
1	9	4	4	0
2	2	6	8	0

(2 marks: 1 mark for using long multiplication with no more than one error; 1 mark for correct answer. Do not award any marks if the 0 for multiplying by a ten is missing.)

14.

				1	8	1
3	2	5	7	9	2	
			3	2		
			2	5	9	
			2	5	6	
					3	2
					3	2
						0

(2 marks: 1 mark for using long division with no more than one error; 1 mark for correct answer)

Test 3

1. 81.5 **(1 mark)**
2. 907 **(1 mark)**
3. 240 **(1 mark)**

4. 200   **(1 mark)**

5. 95 920   **(1 mark)**

6. 5732   **(1 mark)**

7. $1\frac{3}{5}$ (Accept $\frac{8}{5}$)   **(1 mark)**

8. $\frac{71}{100}$ **(1 mark)**

9. 113   **(1 mark)**

10.

		5	0	6
	×		5	2
	1	0	1	2
2	5	3	0	0
2	6	3	1	2

(2 marks: 1 mark for using long multiplication with no more than one error; 1 mark for correct answer. Do not award any marks if the 0 for multiplying by a ten is missing.)

11. $3\frac{1}{8}$ **(1 mark)**

12. $\frac{2}{12}$ (Accept $\frac{1}{6}$)   **(1 mark)**

13.

			2	4		
3	2	7	6	8		
		6	4			
		1	2	8		
		1	2	8		
				0		

(2 marks: 1 mark for using long division with no more than one error; 1 mark for correct answer.)

14. 30

(2 marks: 1 mark for correct order of calculation (subtraction, division, addition); 1 mark for correct answer.)

Test 4

1. $\frac{7}{12}$ **(1 mark)**

2.

	1	3	8	6			1	6	4	3
+		2	5	7		−		4	0	6
	1	6	4	3			1	2	3	7

(2 marks: 1 mark for a correct first calculation, e.g. 1386 + 257 = 1643; 1 mark for correct answer)

3. 141   **(1 mark)**

4. $\mathbf{80} \times 10 + \mathbf{80} \times \mathbf{6}$

(1 mark: both answers needed for 1 mark)

5. 87   **(1 mark)**

6. 846   **(1 mark)**

7. 70   **(1 mark)**

8. $11\frac{19}{20}$ **(1 mark)**

9. $\frac{1}{4}$ (Accept $\frac{6}{24}$)   **(1 mark)**

10. 0.36   **(1 mark)**

11. 55 932   **(1 mark)**

12.

		7	1	5
	×		3	8
	5	7	2	0
2	1	4	5	0
2	7	1	7	0

(2 marks: 1 mark for using long multiplication with no more than one error; 1 mark for correct answer. Do not award any marks if the 0 for multiplying by a ten is missing.)

13. 304 060   **(1 mark)**

14. $\frac{1}{24}$ **(1 mark)**

15. 7.2   **(1 mark)**

Test 5

1. 735 254   **(1 mark)**

2. 35   **(1 mark)**

3. 16 181   **(1 mark)**

4. 6   **(1 mark)**

5. 1000   **(1 mark)**

6. $2\frac{2}{5}$ (Accept $\frac{12}{5}$)   **(1 mark)**

7. 75%   **(1 mark)**

8. 14.25   **(1 mark)**

9. 6.3   **(1 mark)**

10. $\frac{1}{10}$ **(1 mark)**

11.

			2	7	9	.	5
2	8	7	8	2	6	.	0
		5	6				
		2	2	2			
		1	9	6			
			2	6	6		
			2	5	2		
				1	4	0	
				1	4	0	
						0	

(Accept $279\frac{14}{28}$, $279\frac{1}{2}$, 279 r 14)

(2 marks: 1 mark for using long division with no more than one error; 1 mark for correct answer)

12. 10   **(1 mark)**

13.

		5	9	1	8
	×			5	6
	3	5	5	0	8
2	9	5	9	0	0
3	3	1	4	0	8

(2 marks: 1 mark for using long multiplication with no more than one error; 1 mark for correct answer. Do not award any marks if the 0 for multiplying by a ten is missing.)

14. 0.608   **(1 mark)**

15. $5\frac{7}{10}$ **(1 mark)**

Test 6

1. 500 000   **(1 mark)**

2. $\frac{43}{8}$ **(1 mark)**

3. $2\frac{1}{2}$ (Accept $2\frac{3}{6}$, $\frac{15}{6}$)   **(1 mark)**

4. 4985   **(1 mark)**

5. 4498   **(1 mark)**

6. 100 000   **(1 mark)**

7. 107.54   **(1 mark)**

8. 0.076   **(1 mark)**

9. $\frac{1}{10}$ (Accept $\frac{2}{20}$)   **(1 mark)**

10. $11\frac{11}{20}$ **(1 mark)**

11. 62   **(1 mark)**

12.

		7	5	9	3
	×			7	4
	3	0	3	7	2
5	3	1	5	1	0
5	6	1	8	8	2

(2 marks: 1 mark for using long multiplication with no more than one error; 1 mark for correct answer. Do not award any marks if the 0 for multiplying by a ten is missing.)

13.

				8	6	.	7	5
2	4	2	0	8	2			
		1	9	2				
			1	6	2			
			1	4	4			
				1	8	0		
				1	6	8		
					1	2	0	
					1	2	0	
							0	

(Accept $86\frac{18}{24}$, $86\frac{3}{4}$, 86 r 18)

(2 marks: 1 mark for using long division with no more than one error; 1 mark for correct answer.)

Using Mathematics: Answers

Test 1

1. 80 **(1 mark)**
2. 7.20 (Accept 7.20am, 7.20pm, 19.20, 20 past 7) **(1 mark)**
3. 5272   5277   5727   5772   **(1 mark)**
4.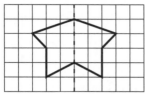
   (Accept lines drawn to within 2mm of vertices. Ignore lines that are not straight.) **(1 mark)**
5. 484 minutes **(1 mark)**
6. 460 305 **(1 mark)**
7. 2 and 29 (Accept answers in either order.) **(1 mark)**
8. $\dfrac{6}{10} = \dfrac{3}{5}$
   (2 marks: 1 mark for each correct fraction)
9. 13.25 (Accept 1.25pm, 25 past 1) **(1 mark)**
10. 150 **(1 mark)**
11. 10 and 5 (Accept answers in either order.) **(1 mark)**
12. a. A cube has **6** faces. **(1 mark)**
    b. A cube has **12** edges. **(1 mark)**
    c. A cube has **8** vertices. **(1 mark)**

Test 2

1. 493 **(1 mark)**
2.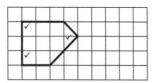
   (2 marks: 2 marks for three angles marked correctly; 1 mark for two correctly marked angles)
3. 5000 **(1 mark)**
4. 1980 **(1 mark)**
5. 3, 6, 15 **(1 mark)**
6. $\dfrac{7}{10}, \dfrac{13}{20}, \dfrac{61}{100}, \dfrac{3}{5}$ (Accept $\dfrac{70}{100}$, $\dfrac{65}{100}, \dfrac{61}{100}, \dfrac{60}{100}$) **(1 mark)**

7. The sides and angles are equal and there are 5 sides. (Accept: The sides and angles in Shape A are equal but they are not in Shape B. Also accept: Shape B is also a pentagon as it has 5 sides, although its sides and/or angles are not all equal.) **(1 mark)**
8. 30cm **(1 mark)**
9. 8°C **(1 mark)**
10. £4.50 (Do not accept £4.5) **(1 mark)**
11. 167, 617, 671, 761
    (2 marks: 1 mark for three correct numbers)
12. 6 **(1 mark)**
13. $\dfrac{80 - \left(\frac{1}{4} \times 80 + 30\right)}{2} = 15$ children

    (2 marks: 1 mark for correct working, but wrong answer)

Test 3

1. $\dfrac{5}{8}$ **(1 mark)**
2. 2 × 8 (Accept 16) **(1 mark)**
3. a. i. 5.2   ii. 5
      (1 mark: Both answers needed for 1 mark.)
   b. i. 14.7   ii. 15
      (1 mark: Both answers needed for 1 mark.)
4. Lines completed as shown.
   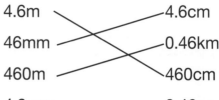
   (2 marks: 1 mark for two or three correctly drawn lines.)
5. 6300 (90 × 70) **(1 mark)**

6. 540 280 **(1 mark)**
7. 180° **(1 mark)**
8. a. 12 boxes (Do not accept
      12 with any remainder,
      decimal or fraction.) **(1 mark)**
   b. 4800 packs **(1 mark)**
9. 4th box ticked only. **(1 mark)**
10. 6cm **(1 mark)**
11. a. Possible answers include:
      24 × 1, 12 × 2, 8 × 3 (Accept
      answers with decimals if
      correct, e.g. 16 × 1.5. Do
      not accept 6 × 4) **(1 mark)**
    b. Possible answers include:
      9 × 1, 8 × 2, 7 × 3, 5 × 5 (Accept
      answers with decimals if correct,
      e.g. 8.5 × 1.5. Do not accept
      6 × 4) **(1 mark)**
12. $\frac{45}{75}, \frac{18}{30}, \frac{36}{60}$ circled only.
      **(2 marks: 1 mark for two fractions
      correctly circled)**

Test 4

1. 2nd and 3rd boxes ticked only.
   (2 marks: 1 mark for either box ticked)
2. 3.09, 3.39, 9.09. 9.93, 39.39 **(1 mark)**
3. 1, 36, 64 circled only.
   **(2 marks: 1 mark for two numbers
   correctly circled)**
4. 55 + 35 + 50 = 130; 130 minutes =
   2 hours 20 minutes
   **(2 marks: 1 mark for correct working,
   but wrong answer)**
5. $\frac{56 - (17 \times 2)}{2} = 11cm$
   **(2 marks: 1 mark for correct working,
   but wrong answer)**
6. $\frac{2}{3}$ **(1 mark)**
7. 15.75 **(1 mark)**
8. $x = 70°$ **(1 mark)**
   $y = 40°$ **(1 mark)**
9. 10 × 3 + 0.1 × 50 = £35
   **(2 marks: 1 mark for correct working,
   but wrong answer)**

Test 5

1. 200 – (40 + 45 + 25 + 30) = 60 minutes
   **(2 marks: 1 mark for correct working,
   but wrong answer)**
2. £12.04 **(1 mark)**
3. 3rd box ticked only. **(1 mark)**
4. 85% **(1 mark)**
5. a. 4512 miles **(1 mark)**
   b. 47 gallons **(1 mark)**
6. 20 ÷ 8 = 2.5, 600 × 2.5 = 1500kg
   **(2 marks: 1 mark for correct working,
   but wrong answer)**
7. 2, 2
   (2 marks: 1 mark for each correct answer)
8. 450ml **(1 mark)**
9. 48 miles **(1 mark)**
10.
   (2 marks: 1 mark for two correct answers)

Test 6

1. 28th (May) **(1 mark)**
2. 18 minutes **(1 mark)**
3. a. 115 706 **(1 mark)**
   b. 250 850 **(1 mark)**
4. 2.903, 2.9, 2.593, 2.33 **(1 mark)**
5. a. 3 **(1 mark)**
   b. 12 **(1 mark)**
6. 64cm **(1 mark)**
7. 270° **(1 mark)**
8. a. 319 500 **(1 mark)**
   b. 300 000 **(1 mark)**
9. 4th box ticked only.
10. □ = 3
    △ = 12 **(1 mark: 1 mark for both
    correct answers)**
11. a. (3,2) marked.
      **(1 mark: 1 mark if the point (3,2)
      has been marked or used as a
      vertex in drawing the square)**
    b. (–2, –1) **(1 mark)**

12. $\frac{15}{24}, \frac{5}{6}$

(2 marks: 1 mark for each correct answer)

Test 7

1. **a.** 29 games **(1 mark)**
   b. 10 games **(1 mark)**
2. 19, 29, 59

(2 marks: 1 mark for two correct answers)

3. 6.50 (Accept 6.50am, 6.50pm, 18.50, 10 to 7) **(1 mark)**
4. $2\frac{1}{6}$

(3 marks: 1 mark for correct addition: $\frac{26}{12}$; 2 marks for correct mixed number: $2\frac{2}{12}$)

5. –5°C **(1 mark)**
6. 52, 46, 34

(2 marks: 1 mark for any two correct numbers correctly placed in the sequence, e.g. 52, 44, 34)

7. $\frac{1}{2}, \frac{7}{12}, \frac{2}{3}, \frac{3}{4}, \frac{5}{6}$

(Accept $\frac{6}{12}, \frac{7}{12}, \frac{8}{12}, \frac{9}{12}, \frac{10}{12}$) **(1 mark)**

8. 6 200 315 **(1 mark)**
9. $x = 100°$ **(1 mark)**
   $y = 80°$ **(1 mark)**
10. **a.** 37.5 **(1 mark)**
   b. 600 **(1 mark)**

Test 8

1. 0.25 litres **(1 mark)**
2. 86 258 – (35 453 + 27 328) = 23 477 saloons

(2 marks: 1 mark for correct working, but wrong answer)

3. **a.** 272 books **(1 mark)**
   b. 6528 books **(1 mark)**
4.

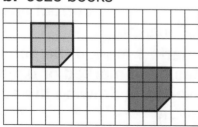

(2 marks: 2 marks for drawing

as shown; 1 mark for correctly orientated and sized shape translated 7 units right or 3 units down)
5. 55° (Accept angle drawn +/– 2°)

(1 mark)

6. 2 ÷ 5 = ticked only. **(1 mark)**
7. 10 – (2.3 + 1.6) = 6.1kg or 10 000 – (2300 + 1600) = 6100g (Accept 6.1kg or 6100g; Units must be correct, e.g. Do not accept 6.1g or 6100kg.)

(2 marks: 1 mark for correct working, but wrong answer)

8. 4 **(1 mark)**
9. Pyramid **(1 mark)**
10. (2365 + 3850) ÷ 24 = 258 boxes (Do not accept 258 with any remainder, decimal or fraction.)

(2 marks: 1 mark for correct working, but wrong answer)

Test 9

1. 192 boxes **(1 mark)**
2. 60, 6

(1 mark: both answers needed for 1 mark)

3.

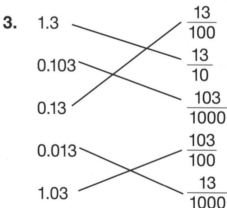

(2 marks: 1 mark for three or four correctly drawn lines)

4. **a.** 3°C **(1 mark)**
   b. 9°C **(1 mark)**
   c. 3 days **(1 mark)**
5. 8 542 135 **(1 mark)**
6. (9 × 9) – (12 × 6) = 81 – 72 = 9cm²

(2 marks: 1 mark for at least one correct calculation of an area i.e. 81 or 72)

7. 13 754 **(1 mark)**
8. **a.** 120 **(1 mark)**

b. 300 **(1 mark)**

c. 25% **(1 mark)**

9. 16 **(1 mark)**

Test 10

1. 59 650 **(1 mark)**

2. 7.324 **(1 mark)**

3. 10, 75, 4

(2 marks: 1 mark for two correct answers)

4.

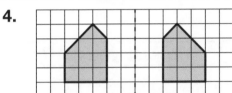

(2 marks: 1 mark for correctly reflected shape in incorrect position)

5. Possible answers:

2 × 6, 3 × 4 in either order. (Also accept 1 × 12 in either order. Do not accept fractions or decimals.) **(1 mark)**

6. **a.** 32 **(1 mark)**

b. 0.047 **(1 mark)**

7. 3rd box ticked only. **(1 mark)**

8. 48, 20 **(1 mark)**

9. $x = 145°$ **(1 mark)**

$y = 35°$ **(1 mark)**

10. **a.** 9cm² **(1 mark)**

b. Square, with 3cm sides (Accept a square with sides of 3cm drawn anywhere on the grid. Accept written answer, e.g. a square with 3cm sides.) **(1 mark)**

Test 11

1. 4th box ticked only. **(1 mark)**

2. 2 634 891 circled only. **(1 mark)**

3. **a.** 2nd box ticked only. **(1 mark)**

b. 3rd box ticked only. **(1 mark)**

4. **(1 mark)**

```
  6 5 1 8
- 3 0 9 4
  ───────
  3 4 2 4
```

5. 20 − (7 + 3.20 + 3.20 + 1.50 + 0.90) = £4.20 (Do not accept £4.2)

(2 marks: 1 mark for correct working, but wrong answer)

6.

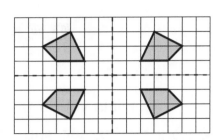

(2 marks: 1 mark for each shape correctly orientated and sized)

7. **a.** 7 apples **(1 mark)**

b. 35 apples **(1 mark)**

8. 5 (Accept correct totals all listed: 5, 6, 7, 8, 9) **(1 mark)**

9. 6, 9

(2 marks: 1 mark for each answer)

10. $x = 60°$ **(1 mark)**

$y = 120°$ **(1 mark)**

Test 12

1. **a.** 8, 9, 16

(1 mark: all answers needed for 1 mark)

b. 10, 24, 20

(1 mark: all answers needed for 1 mark)

2. 8, 4 **(1 mark)**

3. 160cm² **(1 mark)**

4. 20 × 12 × 8 = £1920

(2 marks: 1 mark for correct working, but wrong answer)

5. 29km² +/− 1km² **(1 mark)**

6. 15.4 pounds **(1 mark)**

7. 1056cm² **(1 mark)**

8. **a.** $6\frac{3}{8}$ **(1 mark)**

b. $3\frac{10}{12}$ (Accept $3\frac{5}{6}$) **(1 mark)**

9. −12 **(1 mark)**

10. $A = 12$, $B = 4$

(2 marks: 1 mark for numbers in reverse order)

11. **a.** (−3, −1) (−1, 1) (2, 1) **(1 mark)**

b. Possible answers: (0, −1) (−6, −1) (4, 3) **(1 mark)**

Progress Report

Fill in your total marks for each completed test.

Colour the stars to show how you feel after completing each test.

☆ = needs practice   ☆☆ = nearly there   ☆☆☆ = got it!

Arithmetic

Test	Marks	How do you feel?
Test 1	/ 17	☆ ☆ ☆
Test 2	/ 17	☆ ☆ ☆
Test 3	/ 17	☆ ☆ ☆
Test 4	/ 17	☆ ☆ ☆
Test 5	/ 17	☆ ☆ ☆
Test 6	/ 15	☆ ☆ ☆

Using Mathematics

Test	Marks	How do you feel?
Test 1	/ 15	☆ ☆ ☆
Test 2	/ 16	☆ ☆ ☆
Test 3	/ 17	☆ ☆ ☆
Test 4	/ 15	☆ ☆ ☆
Test 5	/ 15	☆ ☆ ☆
Test 6	/ 17	☆ ☆ ☆
Test 7	/ 17	☆ ☆ ☆
Test 8	/ 15	☆ ☆ ☆
Test 9	/ 15	☆ ☆ ☆
Test 10	/ 15	☆ ☆ ☆
Test 11	/ 16	☆ ☆ ☆
Test 12	/ 16	☆ ☆ ☆